big & SMALL

Original Korean text by Da-jeong Yu
Illustrations by Mi-ye Jeong
Korean edition © Dawoolim

This English edition published by Big & Small in 2015
by arrangement with Dawoolim
English text edited by Joy Cowley
English edition © Big & Small 2015

Distributed in the United States and Canada by
Lerner Publishing Group, Inc.
241 First Avenue North
Minneapolis, MN 55401 U.S. A.
www.lernerbooks.com

ISBN: 978-1-925186-23-9

Printed in the United States of America

The Flow of Water

Written by Da-jeong Yu
Illustrated by Mi-ye Jeong
Edited by Joy Cowley

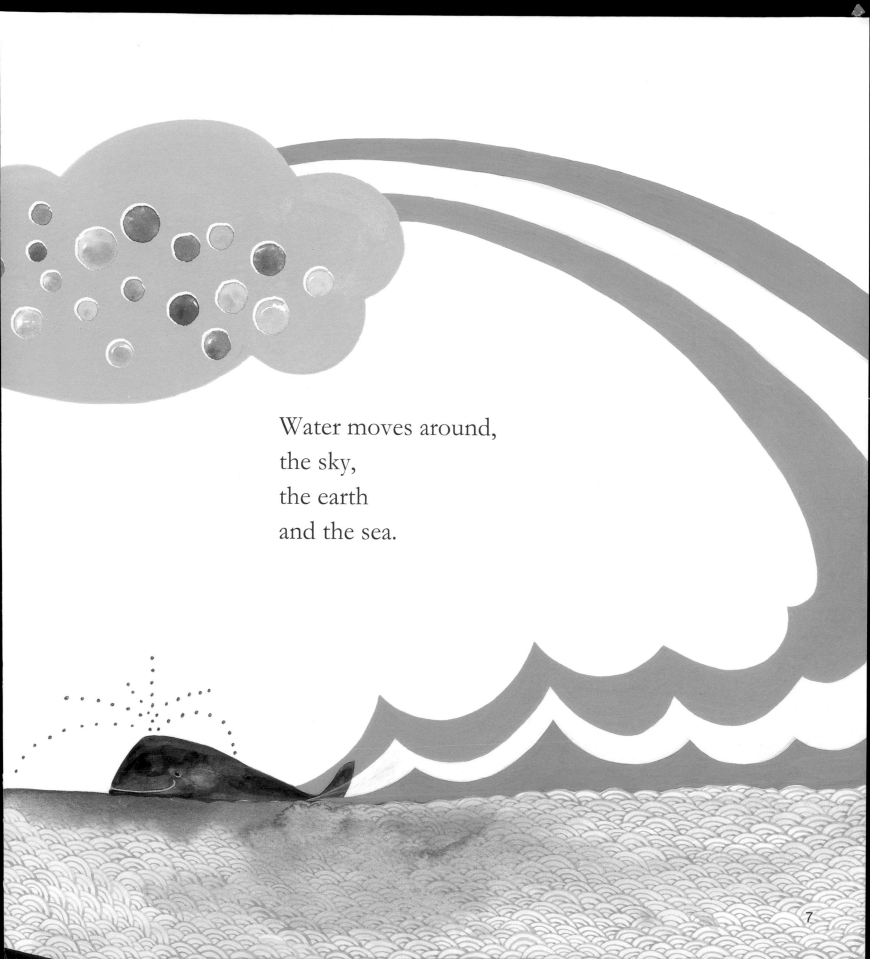

Water moves around,
the sky,
the earth
and the sea.

Early in the morning
water rises as thick mist
on the sea and mountains.

On every green leaf
and on every spider web,

water drops twinkle
as clear and shining dew.

Water becomes fluffy clouds
that float high above flying birds.

The clouds get heavier
and the sky gets darker.
Rain falls from the clouds.
Pitter-patter! Pitter-patter!

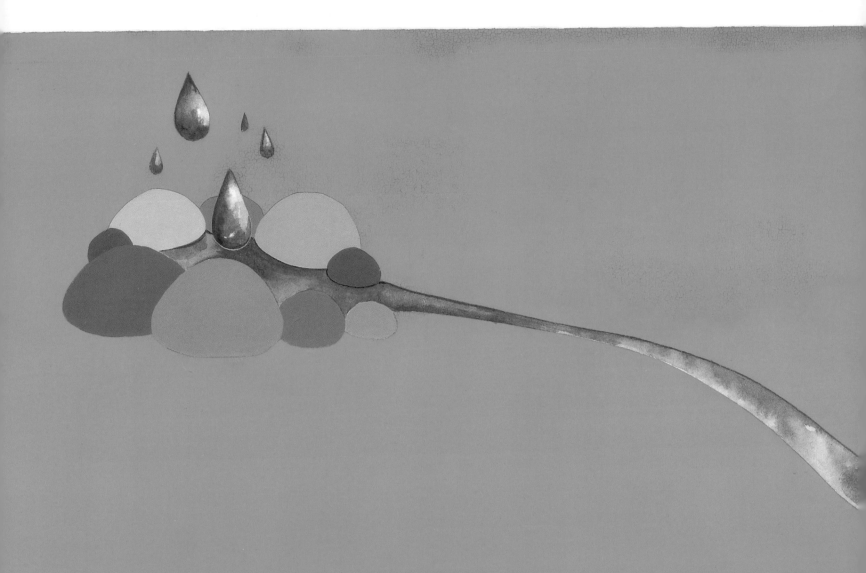

Raindrops come together
to form a flowing trickle
and then a stream.

The stream gets bigger.
It flows along the valley,
through deep forests
and around great rocks.

The stream falls from a cliff
all the way to the bottom.
It splashes, sending up
a great cloud of spray.
It has become waterfall.

In the shadow of a mountain,
the water takes a rest
in a calm and peaceful lake.
Birds fly over it.
Fish dance in it.

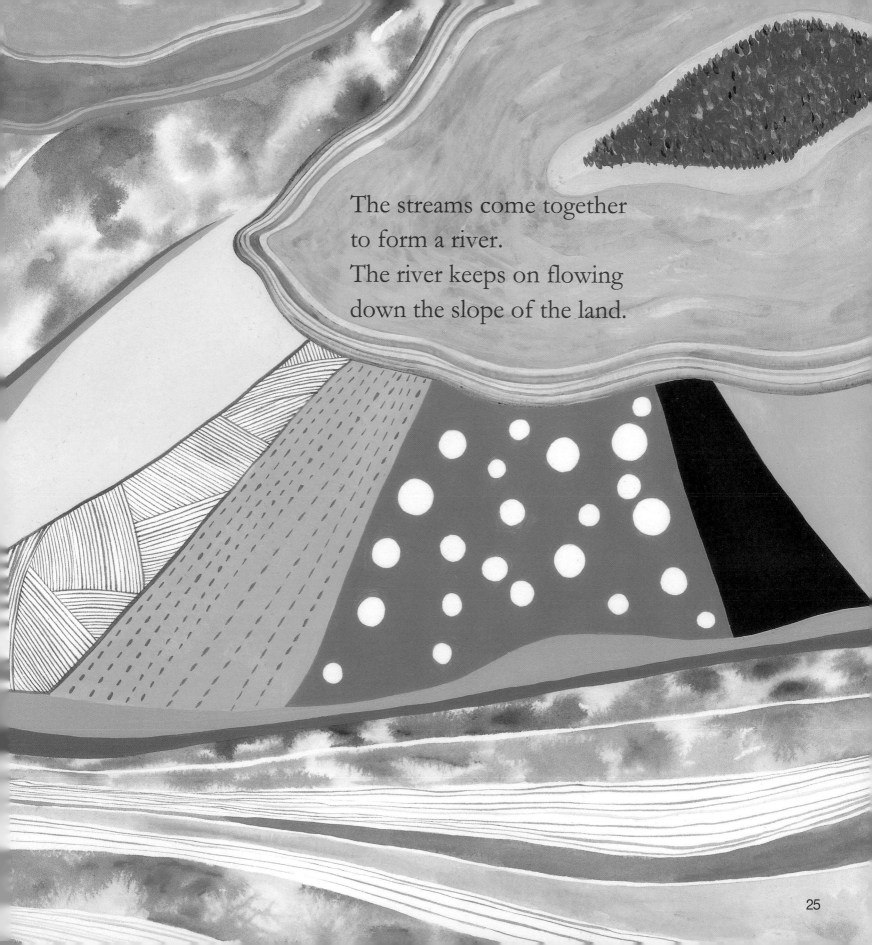

The streams come together
to form a river.
The river keeps on flowing
down the slope of the land.

As the river flows into the sea,
it makes foamy waves.
It has now become a part
of the life of the sea.

As the sea receives
the warm morning sun,
a thick mist rises from it
and the cycle begins again.

The Flow of Water

Water does not stay in one place but continues to travel
while changing its form. This travel is called "water circulation".
Look at water from a tap and then water in nature – dew drops,
a stream, a river, a sea, clouds, rain and snow.

Let's think

How does water change into mist?

How does water change into ice?

Why do rainbows sometimes appear after it rains?

What do you use water for?

Let's do!

Leave a jar outside next time it rains.
After the rain stops, mark the level of rainwater in the jar.
Then leave the same jar of rainwater outside in the sun for a few days.
Check and mark the level of water again.
What happened to the rainwater in the jar?